Niya Georgieva

The World of Fractals

AF135946

Niya Georgieva

The World of Fractals

LAP LAMBERT Academic Publishing

Impressum / Imprint
Bibliografische Information der Deutschen Nationalbibliothek: Die Deutsche Nationalbibliothek verzeichnet diese Publikation in der Deutschen Nationalbibliografie; detaillierte bibliografische Daten sind im Internet über http://dnb.d-nb.de abrufbar.
Alle in diesem Buch genannten Marken und Produktnamen unterliegen warenzeichen-, marken- oder patentrechtlichem Schutz bzw. sind Warenzeichen oder eingetragene Warenzeichen der jeweiligen Inhaber. Die Wiedergabe von Marken, Produktnamen, Gebrauchsnamen, Handelsnamen, Warenbezeichnungen u.s.w. in diesem Werk berechtigt auch ohne besondere Kennzeichnung nicht zu der Annahme, dass solche Namen im Sinne der Warenzeichen- und Markenschutzgesetzgebung als frei zu betrachten wären und daher von jedermann benutzt werden dürften.

Bibliographic information published by the Deutsche Nationalbibliothek: The Deutsche Nationalbibliothek lists this publication in the Deutsche Nationalbibliografie; detailed bibliographic data are available in the Internet at http://dnb.d-nb.de.
Any brand names and product names mentioned in this book are subject to trademark, brand or patent protection and are trademarks or registered trademarks of their respective holders. The use of brand names, product names, common names, trade names, product descriptions etc. even without a particular marking in this works is in no way to be construed to mean that such names may be regarded as unrestricted in respect of trademark and brand protection legislation and could thus be used by anyone.

Coverbild / Cover image: www.ingimage.com

Verlag / Publisher:
LAP LAMBERT Academic Publishing
ist ein Imprint der / is a trademark of
OmniScriptum GmbH & Co. KG
Heinrich-Böcking-Str. 6-8, 66121 Saarbrücken, Deutschland / Germany
Email: info@lap-publishing.com

Herstellung: siehe letzte Seite /
Printed at: see last page
ISBN: 978-3-659-59676-6

Copyright © 2014 OmniScriptum GmbH & Co. KG
Alle Rechte vorbehalten. / All rights reserved. Saarbrücken 2014

Table of contents:

1. Introduction

The world around us abounds in fractals. The branchy systems of trachea tubules, the leaves of the trees, the veins on our hands, river basins, grass, lightning – all these are fractals. Fractals make us reconsider our notions about the geometrical properties

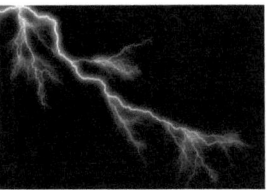

of natural and man-made objects. The underlying reason in studying chaos and fractals is to highlight the objective laws governing all systems which might only appear unpredictable and seemingly chaotic. Such systems are, for instance, cloud formations, weather patterns, turbulent movement of water, migrations of many animals as well as numerous other aspects of life and nature. In all, it will not be too far-fetched to say that the world around us is composed of fractals! What is more, we are gradually becoming aware of them and are starting to recognise them! A number of scientists who have dedicated their lives to studying fractals

have been looking into ways to analyse them and have been trying to find the harmony behind chaos. Thus, peering 'behind the curtain', they have come across the infinite beauty of these formations. Following the tendency in the development of various phenomena and with the aid of new digital technology, they have also laid the basis of a new form of art – fractal art.

2. The world of fractals

A **fractal** is a geometric set, which is radically 'broken' or 'fractured'. The term *fractal* was first used by mathematician Benoit Mandelbrot in 1975. He used this term to describe self-similar patterns which are not clearly differentiable – 'nowhere differentiable'. Later, in 1977, he published **The Fractal Geometry of Nature**. Geometric fractals are known as **determined fractals**. They are also called classical or linear fractals. Here are some of the most fascinating fractals in nature:

Coastal line fractal

A snowflake fractal

2.1. Animals:

Animal behaviour is subject to various attractors; however, inside animals' bodies, as well as on the surface, one can notice many fractal patterns. In the late 1980s, Fujikawa and Matsushita, studying the growth patterns of Bacillus subtilis colonies 168 (B 168), found out that in an environment of insufficient nutrients the colony develops a fractal pattern. Placed under unfavourable conditions fungi colonies, just like the bacteria, grow following fractal patterns. Ammonites, a long extinct group of marine animals, related to the existing species of nautilus, had spiral shells divided into chambers. The walls (septa) of the chambers were connected to the shell through a curved line. In the case of nautilus this is a smooth curve, while in ammonites it is a fractal curve – Koch's curve.

Peacock fractal

Sea shell fractal

Sea-urchin fractal

2.2. Wildlife examples - Leopard spots, zebra stripes

Why and how do spots and stripes appear on the skins of some animals? This coloration most often serves as camouflage and, since it plays a certain role in the survival of the species, it is hereditary genetic information. Leopard spots, for example, are not identical for every individual of the species; however, they are clearly distinctive and different from those of a tiger. What are the mechanisms through which certain species are born with spots, others – with stripes? One of the first models was suggested by Alan Turing, a pioneer in the science of artificial intelligence. Turing's idea was further developed by David Young on the basis of the concept of cellular automata.

2.3. Plants
Fractal structures are especially conspicuous in plants. Their growth is subject to the iterations of the cyclic attractor. Most plants have a branchy shape: the main stem is divided into a number of branches. Each of these is in turn divided into smaller branches and so forth. A tree branch is similar in its structure to the whole tree, and a fern twig looks almost identical to the whole plant.

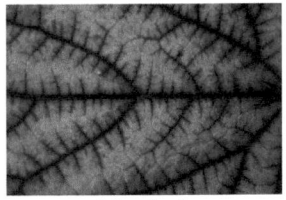  

A tree leaf	Fractal flower	Broccoli fractals

Fractals are self-similar patterns, which have no exact dimension. The branchy systems of trachea tubules, the leaves of the trees, the veins on our hands, rivers, grass, lightning – all these are fractals. Fractals make us reconsider our concepts and notions about the geometrical properties of natural and man-made objects. The underlying reason in studying chaos and fractals is to highlight the objective laws governing all systems which might only appear unpredictable and seemingly chaotic. Such systems are, for instance, cloud formations, weather patterns, turbulent movement of water, migratory behaviour of animals as well as numerous other aspects of life and nature.

It is unthinkable to talk about fractals while ignoring the dynamic processes which have formed them. However, by assuming this approach we might end up in deep waters. What are these processes and what is the mathematical connection that links them? Isn't it also true that complex structures and patterns in nature come as a result of no less complex processes?

The paradigm "The complexity of structures is a result of complex and sophisticated processes" might be true in some cases, but is far from being true on principle. It seems that the opposite is closer to the truth: it is a very simple creative process that underlies a very complex model. This means that if we think of a process as a simple one, we might be misled to infer that we will be able to easily understand the consequences of its implementing.

3. Fractals in geometry

3.1. The principle of reverse connection

One of the best examples of a simple process with a complex development is the process defined by a quadratic equation of the type x^2+c, where c is a constant (in some cases it can be a non-fixed constant as is with Mandelbrot sets). The reverse connection processes are fundamental in ALL sciences. In fact, they were introduced by Sir Isaac Newton and Gottfried Leibniz about 300 years ago under the form of the laws of dynamics; today they are a standard procedure to model natural phenomena and to apply the laws.

These laws help us determine the position and the velocity of a particle at a particular moment taking the data of a preceding moment. Then the movement of the particle can be described by iteration. It does not matter whether the processes are discrete (processes seen as a sequence of discrete phases) or continuous.

By using the quadratic equation of the type x^2+c and completing numerous iterations and modifications we have come to some of the most popular and widespread fractal constructions.

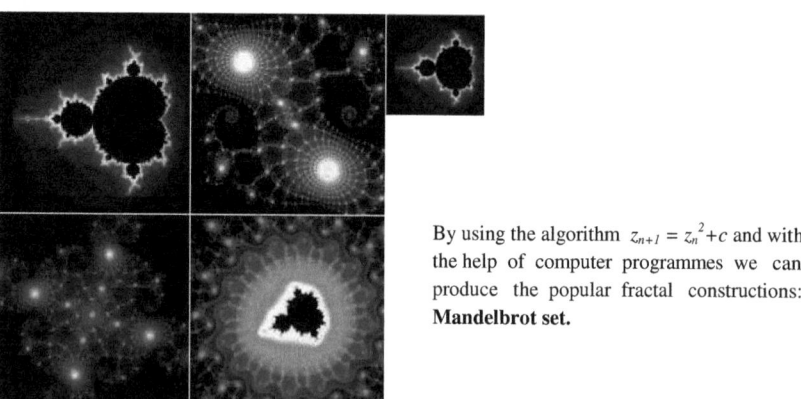

By using the algorithm $z_{n+1} = z_n^2 + c$ and with the help of computer programmes we can produce the popular fractal constructions: **Mandelbrot set.**

These complex fractal constructions remind us of some of the well-known geometrical figures.

The **cardioid** is well-known from many fractal patterns. It is the cardioid that is the first and the largest element in Mandelbrot sets. In geometry the cardioid is a plane curve of fouth grade, a type of epicycloid with a single cusp.

The cardioid is a curve which is traced by a point on the perimeter of a circle with a radius *a* that is rolling around a fixed circle of the same radius. It is also a special type of Pascal snail (limacon) with a single cusp, which occurs when the ratio between the radiuses of the two circles is **1**. A third definition for this curve is as an inverse curve of a parabola.

The name of the cardioid comes from Greek: καρδια**"heart"** + ειδος**"shape"**. The discovery of the curve is said to belong to Dutch mathematician Keursma in the late 17th century, and the name was first used by the Italian mathematician de Castillion in 1741 in his work "De curva cardioïde".

The concept of the iteration of identical structures, which form figures that are self-similar to the initial figures, can be illustrated in math tasks for young children. The process of learning starts with the basics of the complex processes. On this basis, it becomes easier to conceive the complexity in any process.

Problem 1.

The figure below consists of a succession of attached "houses". In each "house" the area of the roof is **4** times smaller than the area of the square. The area of the square of each new "house" is twice as large as the area of the roof of the preceding "house".

A) If the area of the initial "house" (the black one) is **80 square cm**, what is the side of the square of the initial "house" and the area of the fouth "house"?

B) If the area of the initial "house" (the black one) is **1280 squre cm**, how many times is the area of the ninth "house" smaller than the area of the initial "house"?

This math problem for younger students demonstrates the method of constructing a Pythagorean tree, which we will discuss later.

Solution:

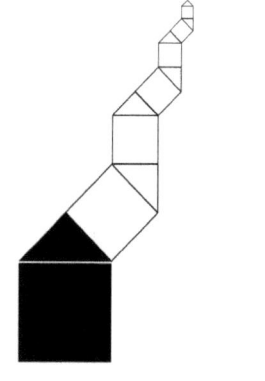

A)

$S_1 = 80 \text{ cm}^2;$ $\quad$ $4x = S_{\text{triangle 1}},$ $\quad$ $S_{\text{square 1}} = 16x$

$S_1 = 20x = 80 \text{ cm}^2 \Rightarrow x = 4 \text{ cm};$

$S_{\text{house 2}} = S_{\text{square 2}} + S_{\text{triangle 2}} = 2.4x + 2x = 10x;$

$S_{\text{house 3}} = S_{\text{square 3}} + S_{\text{triangle 3}} = 2.2x + x = 5x;$

$S_1 = 80 \text{ cm}^2 = S_{\text{triangle 1}} + S_{\text{square 1}} = 5.S_{\text{triangle 1}}$

$\Rightarrow S_{\text{square 1}} = 64 \text{ cm}^2 \Rightarrow$ **Side of the square = 8cm.**

B)

$S_1 = 1280 \text{ cm}^2;$

S_1 is 2^9 times larger than S_9, since each new "house" has an area **2** times smaller than the preceding one. The houses are similar with a similarity factor equal to $\sqrt{2}$.

Problem 2.

For **12** days, **75** bulls can graze the grass on a meadow with an area of **60** acres, as well as the grass which has grown again on the meadow for the same period of **12** days. For **15** days, **81** bulls can graze the grass on a meadow with an area of **72** acres, as well as the newly-grown grass for the same period of **15** days. How many bulls can, for **18** days, graze the grass on a meadow with an area of **96** acres, together with the grass which has grown on the meadow for this period of **18** days?

Solution:

Let's use:

a - for the initial amount of grass on **1** acre;

b – for the amount of grass a bull can graze for **1** day;

c – for the amount of grass on **1** acre which has grown again for **1** day;

x – for the number of bulls which for **18** days will graze the **96** acres together with the newly-grown grass for this period.

$$60(a+12c) = 75.12.b$$
$$72(a+15c) = 71.15.b$$
$$96(a+18c) = x.18.b$$

The solution is: **x=100** bulls

Problem 3. The triangle numbers and the square numbers can be represented in the following way:

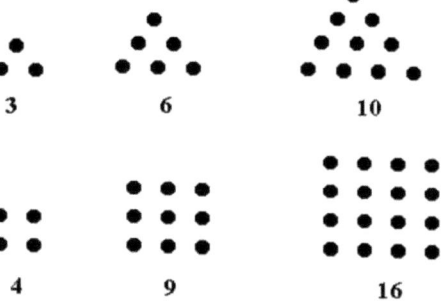

Find two successive triangle numbers whose difference equals to 11 and two successive square numbers whose difference also equals to 11. What is the sum of these four numbers?

Triangle numbers consisting of:

   1) 3 points (+3)

   2) 6 points (+4)

   3) 10 points (+5)

   4) 15 points (+6)

   5) 21 points (+7)

   6) 28 points (+8)

   7) 36 points (+9)

   8) 45 points (+10)

   9) 55 points (+11)

   10) 66 points ; 66+55=121

Square numbers consisting of:

   1) 4 points (+5)

   2) 9 points (+7)

   3) 16 points (+9)

   4) 25 points (+11)

   5) 36 points ; 25+36=61

⇨ 121+61=182.

4. Fractal dimensions and methods of calculation

The dimension is a ratio which shows the amount of information needed to describe a particular object. If we consider the shape, size and the position of elemental geometrical figures, we can get an idea of some popular fractals and their dimensions. Let's start with the simplest method of calculating fractal dimension which uses the property of self-similarity.

4.1. The method of self-similarity

Let's take a square – it is a two-dimensional plane figure. If we enlarge it to an area twice as large, the result will be a new figure, which consists of four identical squares and is self-similar to the original one. If we do the same with a triangle, another two-dimensional plane figure, we will get four identical triangles composing a figure which is self-similar to the original one.

Now, let's take a cube. It is a three-dimensional figure. If we make it twice as big, we will have eight identical cubes as a result.

These three examples show a clearly outlined regularity.

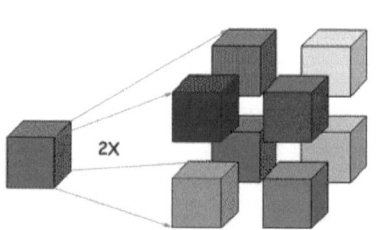

If we raise the enlargement of the figure **e** to power, where the exponent is equa to the dimension **D**, we will invariably get the number of

the resultant figures:

$$N: e^D = N.$$

Therefore, with the help of this formula we can find out the dimension **D**:

D = log N / log e

Koch's curve is considered to be one of the most typical geometric fractals. The algorithm of its creation is described as follows:

The line segment is divided into three segments of equal length. An equilateral triangle is drawn that has the middle segment from step **1** as its base and points outward. The segment that is the base of the triangle is then removed. The process is iterated with every new line segment.

Now, let's consider **Koch's curve** after a few iterations of this process. We can identify four identical snowflakes in it (**N = 4**). Each of them equals to **1/3** of each shape, so **e = 3**. Applying the formula, we find out the dimension: **D = log 4 / log 3 =1.26**.

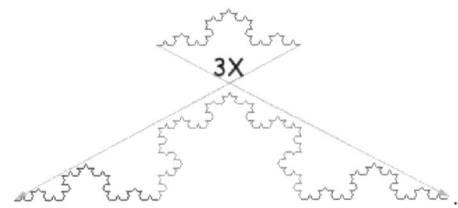

There are other approaches to calculating fractal dimensions.

4.2. Geometrical method

The similarity method is very effective in calculating fractal dimension of fractals made up of a set of self-identical versions. However, if we try to apply it to the coastline of England, it becomes an impossible task. The problem is that in this case all lines have different sizes and require different magnifications. There is a simple way to tackle this.

As we know, a natural fractal has an infinite number of details. This means that additional details should be taken into consideration in its magnification which are added up to its size. In non-fractal figures, however, the size never changes.

Let's take the chart below as an example. It shows the sizes of some non-fractals under different magnification. If we draw a chart of the logarithm of size above the logarithm of magnification, we will get horizontal lines. This demonstrates that the sizes do not change, which means that the figures are simply not fractals.

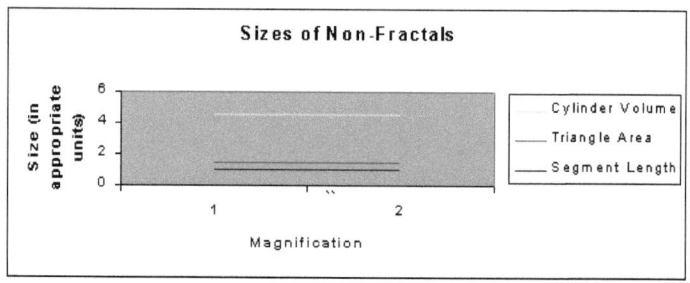

Now we take a few fractals and do the same. What we get is not horizontal lines since sizes increase with the increase in magnification. This is a proof that the figures are fractals.

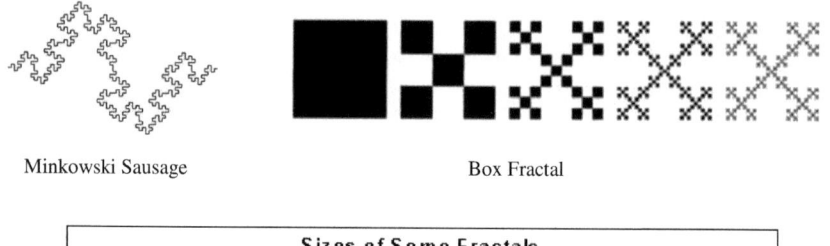

Minkowski Sausage   Box Fractal

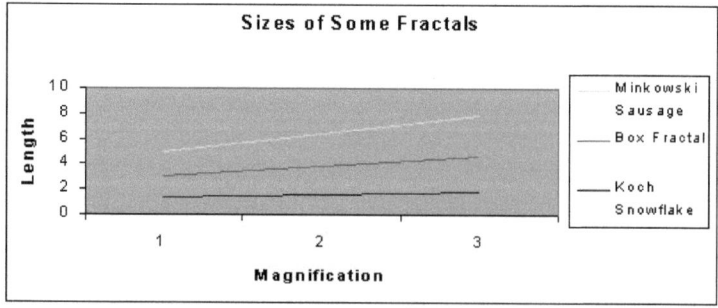

Now we can easily calculate fractal dimensions using the inclinations (slopes) of the lines in a simple formula:

Fractal dimension = inclination + 1

12

The geometrical method can be used very effectively for natural irregular formations with Brownian similarity (it works despite the random and chaotic dispersal of the elements). It is also used in calculations of the dimension of coastlines, borders and cloud formations.

4.3. Box-counting method

The similarity method and the geometrical method for calculating fractal dimensions require measuring the size. In many fractals, however, this is not only very complex, it is often impossible. For example, take the fractal on the right.

A simpler method in such cases is box counting. We place the fractal on a sheet of grid paper, each square with a side **h**. Then we count the squares which are not empty. Their number is to be denoted with the variable **N.** If we use a grid with smaller square sides, we get a more accurate calculation, which is, in practice, magnification. In fact, the magnification is equal to **1/h**. When we described the similarity method for calculating fractal dimension, we used the following formula: **D = log N / log e**. Now, we can replace it with: **D = log N / log (1/h)**

If we make **h** smaller, that is, if we use a finer grid, we can define the dimension even more accurately. For three-dimensional fractals cubes will be used instead of squares; for one-dimensional – segments.

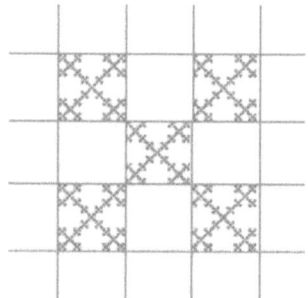
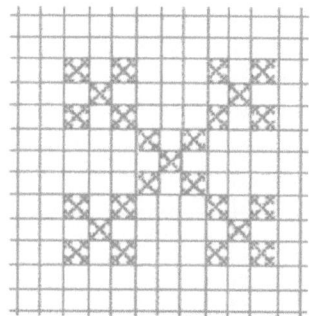

For example, let's calculate the fractal dimensions of Box fractal. We place it on a grid paper with squares measuring **1/3** and **1/9**. As you can see, in the first grid **5** squres are covered; in the second - **25**.
In the first case:

D=log5/log(1/(1/3))=1.46.

In the second case the answer is the same, which means that the calculated dimension is correct.
This method is very effective for natural fractal formations, which are difficult to measure, for example, bacteria clusters.
Trees and ferns are natural fractals which can be computer-modeled with recursive algorithms. Their recursive nature is demonstrated with the following example: take a tree branch or a fern leaf and you will see that they are miniature copies of the whole plant, not identical but similar in essence.
Another simple example is Cantor sets, in which increasingly lesser open middle segments are deleted from the line segment **[0, 1]**, resulting in a set which can (or cannot) be self-similar while magnified and can (or cannot) have dimension **d**, for which **0 < d < 1**. A simple rule, such as removing the number **7** from decimal fractions generates a set, which is self-similar at tenfold magnification and has a dimension of **log9/log10** (the value does not chance irrespective of the logarithm base) and thus demonstrates the connection between the two concepts.
Its fractal dimension is: **ln2/ln3 = 0.63** – Cantor comb.

5. Popular fractal examples

Koch's curve is considered to be one of the most typical geometric fractals. It was named after Helge von Koch and is one of the earliest fractals to be described by the

14

famous Swedish mathematician in 1904. The curved line (the generator) of the fractal is an equilateral triangle whose sides equal to 1/3 of the length of the tangent.

Koch's snowflakes

The algorithm is described as follows:

- The line segment is divided into three segments of equal length;
- An equilateral triangle is drawn that has the middle segment from step 1 as its base and points outward;
- The segment that is the base of the triangle is then removed.

The process is iterated with every new line segment.

Now, let's consider another example with **Koch's curve.**

In it we can identify four identical snowflakes (**N = 4**). Each of them equals to **1/3** of each shape, so **e = 3**. Applying the formula, we find out the dimension: **D = log 4 / log 3 =1.26**. Here, the dimension is a fraction, which is something totally untypical of traditional Euclidian geometry.

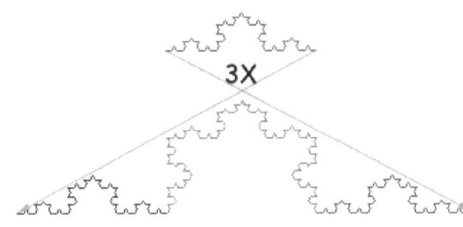

With many fractals self-similarity is obvious. All of them consist of miniature versions of themselves. If they are magnified, it becomes obvious that every version is identical to the whole image. These are what we call geometrical fractals.

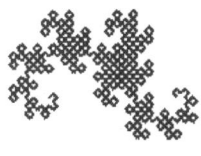

Dragon curve

Durer's pentagon

Levi's curve

Sierpinski triangle

Mandelbrot experimented a lot with Koch's curves and the results of those experiments are shapes such as Koch's islands, Koch's crosses, Koch's snowflakes and even three-dimensional constructions of Koch's curves.

Koch's snowflakes

The algorithm is described as follows:

- The line segment is divided into three segments of equal length;
- An equilateral triangle is drawn that has the middle segment as its base;
- The segment that is the base of the triangle is then removed.

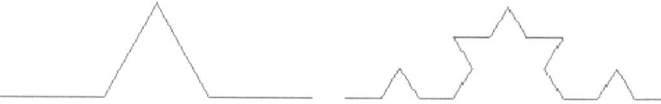

The process is iterated with every new line segment.

Koch's curve is infinite in length. Its area equals to 8/5 of the area of the initial triangle. The infinite perimeter encloses a limited area.

Koch's Crosses

The algorithm is described as follows:

- The line segment is divided into two segments of equal length;
- A short perpendicular segment is drawn in the middle of the line segment.

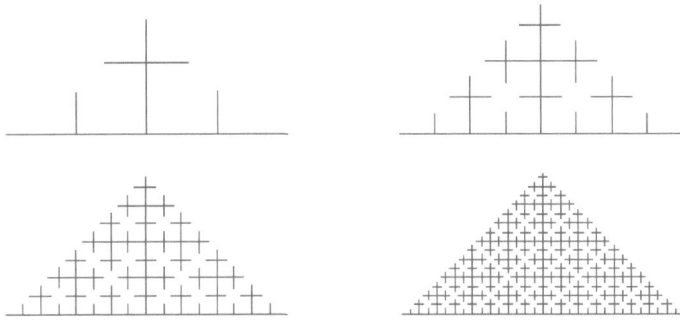

Koch's Islands

The algorithm is described as follows:

- The line segment is divided into three segments, two of which are equal in length, while the third one is smaller;
- An isosceles triangle is drawn that has the middle (smallest) segment as its base;

- The segment that is the base of the triangle is then removed

It is important to point out the three characteristics of fractals: their self-similarity (all parts are similar to the whole fractal) of the shapes at any magnification; the infinite character of fractals; all elements have a context, i.e. every element is part of a larger and complete structure.

The concept of self-similar curves was developed by Paul Pierre Levy. In 1938 he published Plane or Space Curves and Surfaces Consisting of Parts Similar to the Whole, where he describes two fractal curves – **Levy C-curve and Levy Dragon curve**.

Levy C-curve

Levy C-curve is a fractal, whose construction of this fractal starts with a square divided into two parts; then each side is replaced by the same fragment. The infinite iteration of this procedure leads to this curve.

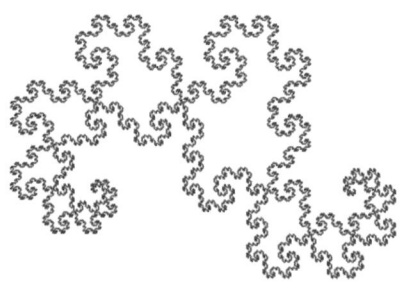

Dragon Curve

And now, a math problem related to the Dragon curve (Jurasic Park Fractal) :

Problem 4.

Dragon: This one is a very interesting curve. It has the outlines of a dragon. People who claim to have seen a dragon say that dragons look like this one.

Now, let's take a long rectangular paper strip whose left end has been marked with a dot. Then we fold the strip into two so that we cover the dot and continue folding it in the same way (the right end over the left end).

After that, we open the strip and place it so that the end marked with the dot is on the left. The folding follows this pattern: down-turn, down-turn, up-turn – DDU. If we fold the strip 3 times and then unfold it, we will see that the turns follow the pattern: DDUDDUU.

A) What is the pattern of turns if the strip is folded 5 times?

B) Describe the shape of the strip after the 6th folding.

Represented on cellular paper, the strip looks as follows:

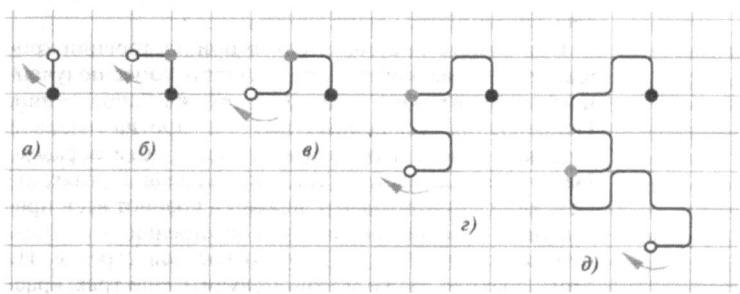

Solution:

A)   1) The number of turns is always odd. If after several foldings their number is k, then after the next folding the number of turns will be 2.k+1:

After 2 foldings: 2.1+1=3

After 3 foldings: 2.3+1=7

After 4 foldings: 2.7+1=15

After 5 foldings: 2.15+1=31

2) In the middle of the strip there is always a down-turn (D), and the turns until the middle D are the same as with the previous folding. With every new move (folding) we witness an additional left down-turn (D) and a right up-turn (U).

3) The turns on the left and on the right of the middle D are always asymmetrical.

Therefore, after the 5th folding, the turns follow this pattern:

DDUDDUUUDDUUDUUDDDUDDUUUDDUUDUU

B) On a piece of cellular paper we replace a D – turn on the left and a U-turn on the right at +90 degrees. Each following folding adds the previous curve rotated at -90 degrees around the last point of the previous move.

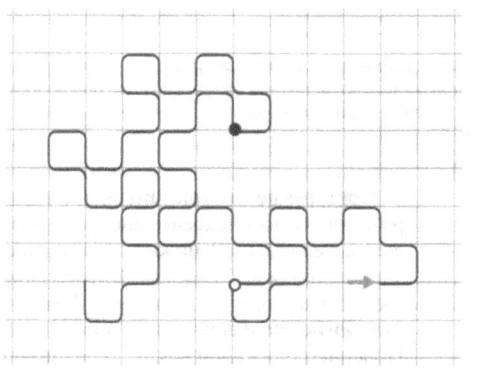

The Sierpinski Triangle

The repetitive nature of this shape follows the algorithm:

Start with a triangle in a plane. The triangle is equilateral.

1. Shrink the triangle to ½ height and ½ width, make three copies, and position the three triangles so that each triangle touches the two other triangles at a corner.
2. Repeat step 1 with each of the smaller triangles.

We can prove that the area of Sierpinski triangle is **0**.

Let's assume that the area of the initial triangle is 1. In step 1 we remove ¼ of its area. In step 2 we remove three triangles, the area of each of which is equal to $(¼)^2$ of the area of the initial triangle; analogically thinking, we can conclude that the overall area of all removed parts is equal to: $¼ + 3. (¼)^2 + 3^2. (¼)^3 + … + 3^{n-1}. (¼)^n + …$ i.e.

$= 1$, which can be easily proved by the ratio $\frac{1}{x-1} = 1 + x + x^2 + x^3 + …$ for $| x |<1$.

Sierpinski triangle has a fractional dimension, which is approximately equal to **1,585**.

Problem 5. The figures below show a row of equilateral triangles with an area of 1. The apexes of the white triangle in fig.2 lie in the middle of the three sides of the initial triangle. If the succession continues, as it is shown in fig.3, what is the total area of the dark triangles in fig.5?

In fig. 2, the white area is ¼ of the total area.

In fig. 3, the white area is $\frac{1}{4} + 3.\left(\frac{1}{4}\right)^2 = \frac{7}{16}$

In fig. 4, the white area is $\frac{7}{16} + 9.\left(\frac{1}{4}\right)^3 = \frac{37}{64}$

In fig. 5, the white area is $\frac{37}{64} + 27.\left(\frac{1}{4}\right)^4 = \frac{175}{256}$

⇨ The dark area in fig. 5 is $\frac{81}{256}$.

Sierpinski triangle can be obtained from Paskal's triangle, through the sequence 1; 1, 1; 1, 0, 1; 1, 1, 1, 1; 1, 0, 0, 0, 1; ... Colouring all odd numbers in Paskal's triangle in black and all even numbers – in white, we obtain Sierpinski triangle.

The above-described infinite iterative algorithm does not depend on the shape of the initial figure. The following images show Sierpinski triangle constructed from a square as an initial figure.

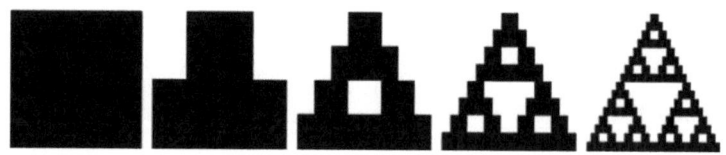

Sierpinski rectangular – Sierpinski carpet

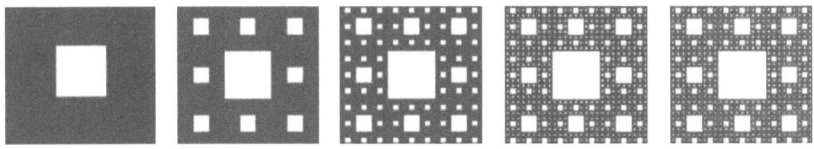

This is Sierpinski triangle curve

The principle of constructing these famous fractals is simple: every element is replaced by N number of elements similar to it. According to their dimensions they can be:

One-dimensional

Or **Cantor comb** – described by Geor Cantor, one of the creators of the set theory, in 1883. The fractal is constructed by removal of the middle third of a segment. The infinite iteration of the operation leads to the development of the so called Cantor dust, a set of points lying on a single line segment, the sum of the lengths of which is equal to 0.

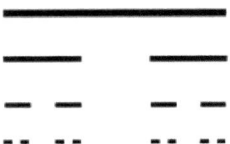

Two-dimentional

These were first described in *"Mosaics made up from pentagons"* in *The Painter's Manual* (1525) by Albrecht Durer. The so called Albrecht Durer's pentagon is obtained by constructing pentagons on the sides of a regular pentagon, which leads to a similar but larger pentagon.

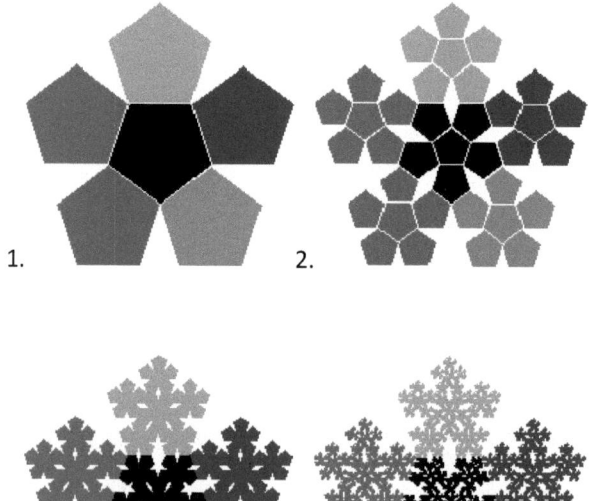

1.   2.

3.   4.

The most famous among three-dimensional types is Sierpinski carpet (1916).

Its dimension is ln20/ln3 ~ 2.7, and it resembles bone structure. After infinite iterations this curve will turn into Cantor dust; the same happens to our bones, by the way.

6. Fractal Modelling

Paul Levy – 1938 (French mathematician)

Problem 6.

Start with an isosceles right triangle L_0. Replace this triangle with two isosceles right triangles so that the hypotenuse of each new triangle lies on one of the equal sides of the old triangle. Place each new triangle so that it points out from the original triangle to get L_1. For the next step, repeat the process on each of the triangles that make up L_1:

A) Iterate the above procedure **4** times;
B) Compare the area of L_0 with the area of L_4;

Solution:

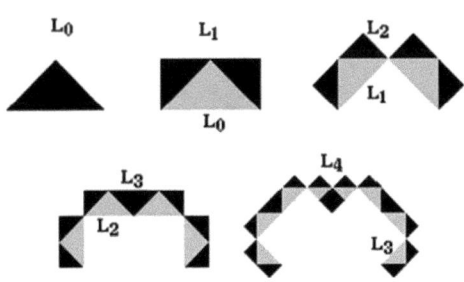

The result after four iterations is the Levy Dragon.

For each new iteration, we replace each triangle in L_k by two isosceles right triangles. The Lévy dragon is the limiting set of this iterative construction. Notice that at each iteration, the sides of each triangle in L_k are scaled by a factor $r = \frac{1}{\sqrt{2}}$, so that the area is scaled by 1/2. But each triangle in L_k produces two new triangles, so the total area remains unchanged. Thus the area of the Lévy dragon is the same as the area of the original isosceles right triangle L_0.

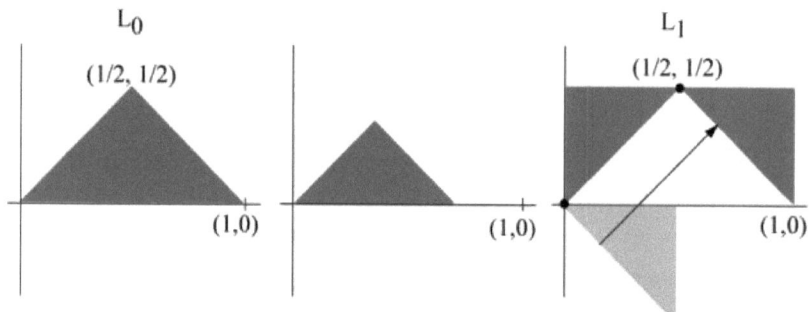

One triangle must then be rotated by **45°**, while the other triangle must be rotated by **-45°** (i.e. in the clockwise direction) and translated by **1/2** in both the x and y directions. This yields iterative functions which reflect the scaling $(r = \frac{1}{\sqrt{2}})$ and the rotation (**by +45°/-45°**).

The attractor (the geometric construct) will be the Lévy Dragon. The Lévy Dragon (the Dragon curve of Levy) consists of two self-similar pieces corresponding to the two functions in the iterated function system.

Levy Curve

26

First four iterations (repetitions) of Levy system

The line segments generated by the **L**-system (Levy system) correspond to the hypotenuses of the isosceles right triangles generated in the construction described above.

Similarity dimension (Self-similarity)

The Levy dragon is self-similar with 2 non-overlapping copies of itself, each scaled by the factor **R < 1**. Therefore the similarity dimension **D**, of the attractor of the iterated function system is the solution to:

$\sum_{k=1}^{2} r^d \Rightarrow D = \log(1/2)/\log(1/\sqrt{2}) = 2$

Special properties/ Specific properties

Paul Lévy studied the curve now known as Lévy's Dragon as part of a more general study of curves consisting of parts similar to the whole. In this study he was motivated by the earlier work of Helge von Koch and the Koch curve.

Among several of the properties that Lévy observed was that the plane can be tiled by copies of the Lévy dragon. This means that there is a sequence of sets congruent to the Lévy dragon that are non-overlapping and whose unions is the entire plane.

Another property shown by Lévy is that the dragon has non-empty interior. Levy curve has fractional dimension of approximately **1.934007183**. Bailey, Kim, and Strichartz show that the interior of the Lévy dragon consists of a countable number of components, the largest of which is a hexagon that is only a speck on the dragon. Moreover, they found **16** different shapes for these components and inferred that there were no others. Alster proved that the number of shapes is finite.

The copy of the Lévy dragon (below) lies on top of a grid of size **1/4 by 1/4** with the initial line segment going from the origin to the point (**1,0**). It demonstrates that the Lévy dragon lies within a rectangle with $-1/2 \leq x \leq 3/2$ and $-1/4 \leq y \leq 1$, i.e. a rectangle of width **2** and height **1.25**

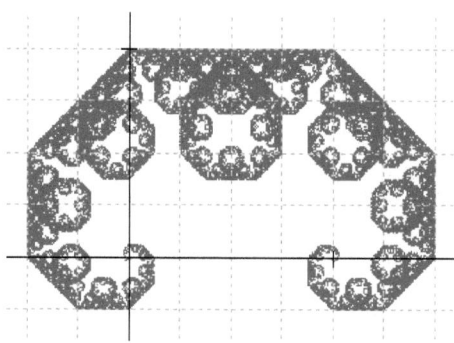

The Lévy dragon can be constructed by replacing a line segment with two segments at **45°**. If the angle between the line segments is less than **45°** then a different dragon curve will be formed. If we let the angle grow from **0°** to **45°**, we can watch the Lévy dragon being born.

The construction of the Lévy dragon and the Heighway dragon (the dragon curve we are going to discuss later) are very similar. In each case one can start with an isosceles right triangle and replace this triangle with two isosceles right triangles so that the hypotenuse of each new triangle lies on one of the equal sides of the old triangle. The difference is how those new triangles are placed relative to sides of the old triangle. For the Lévy dragon, both are placed towards the "outside"; for the Heighway dragon, one is placed pointing out while the other is placed pointing in. Because of this similarity, it is perhaps not surprising that one can transform the Lévy dragon into the Heighway dragon through a continuous transformation. In fact, for each **t** in the interval **[0,1]** an iterated function can be defined.

Let **A(t)** be the unique attractor of the iterated function system corresponding to the value **t**. Then **A** is a continuous function from the unit interval to the space of compact sets with the Hausdorff topology, with **A(0)** equal to the Lévy dragon and **A(1)** equal to the Heighway dragon.

Variations/ Derivative fractals
Levy Tapestry

Rather than start with just one horizontal line segment, one can start the construction of the Lévy Dragon with a square. The iteration steps are repeated for each of the four sides of the square. There are two choices as one can orient the initial triangles to point inside the square or to point outside the square. The resulting images are sometimes called a Lévy Tapestry.

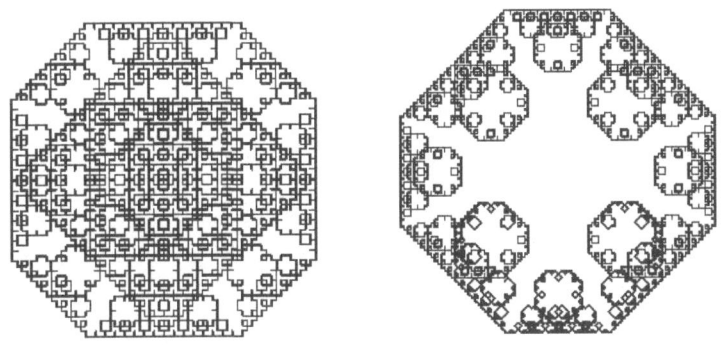

Levy Diamonds

Problem 7.

Start with a trapezoid C_0 with a base of length **1** and the other three sides of length **1/2**. Replace this trapezoid with three scaled copies whose bases lie on the equal sides of the old trapezoid. Place each new trapezoid so that it points out from the original trapezoid to get C_1. For the next step, repeat the process on each of the trapezoids that make up **C1** to get **C2.**

Solution:

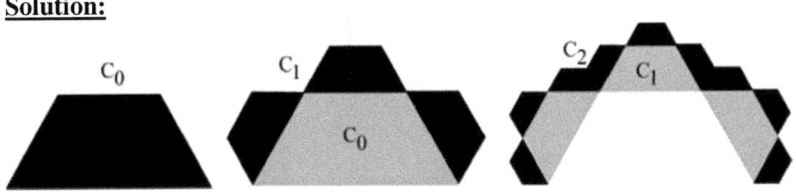

Repeating this construction infinitely yields the following figure that looks like it is formed of many diamond shapes of smaller and smaller scale.

McWorter's Pentigree

Problem 8.

Starting from the beginning of the coordination system **(0,0)**, draw a line segment of length **r** at an angle of **36°**. Turn left by **72°** and draw another line segment of length **r**. Next turn right by **144°** and draw a third line segment of length **r**. The remaining three line segments, each of length **r**, are formed by making a right turn of **72°** followed by two left turns of **72°**. The end of the last line segment will be at the point **(1,0)**.

Solution:

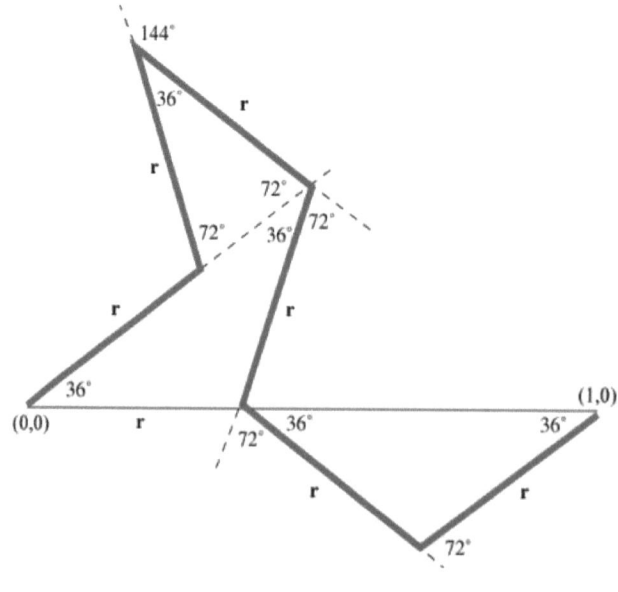

The six red line segments in the figure give the basic motif (first iteration) for this iterative function. Each subsequent iteration is done by replacing each line segment with a scaled version of the basic motif.

These six transformations result from rotations of **36°, 108°, -36°, -108°, -36°**, and **36°** respectively. Thus, the endpoints of the fourth segment and the last segment lie on the x-axis. The value of **R** must be chosen so that the endpoint of the last segment is at the point **(1,0)**. From the figure we see that we must have:

$$r + 2r\cos36° = 1 \Rightarrow r = 1/(1+2\cos36°) = 1/\{1+2[(1+\sqrt{5})/4]\} = (3-\sqrt{5})/2 = 0.381966$$

It is now possible to determine how each of the six rotated segments must be translated. For example, the second segment must be translated to (**r cos36°, r sin36°**) after a rotation by **108°** and a scaling by a factor of **r = 0.381966**. Similar calculations with the other segments lead to the following iterated functions.

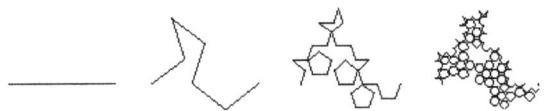

First three iterations

Similarity Dimension (Self-similarity)

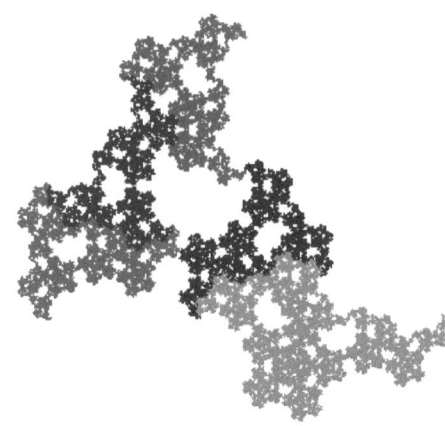

The pentrigree is self-similar with **6** non-overlapping copies of itself, each scaled by the factor **r < 1**. Therefore the similarity dimension **d**, of the attractor of the iterated function is the solution to:

$$\sum_{k=1}^{6} r^d \Rightarrow D = \log(1/6)/\log(r) = 1.86172$$

Special Properties/ Specific Properties

Five copies of the pentigree fit together to form a set with a fivefold rotational symmetry as illustrated in the figure below. This figure is called "the 2nd form of McWorter's pentigree".

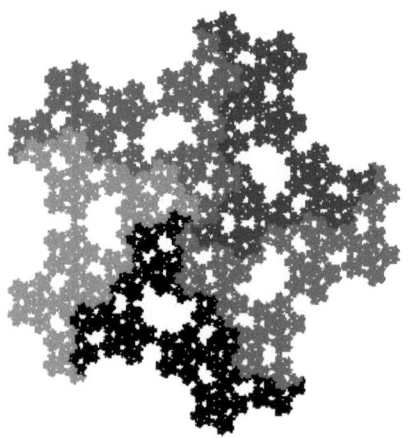

Variations/ Derivative Fractals

The pentadentrite is a variation of McWorter's pentigree based on a different initial angle and scaling factor.

William A. McWorter (1932 – 2009) was an associate professor of mathematics at Ohio State. The pentigree was discovered by accident during his hunts for "dragons", a category of fractals and space-filling curves that McWorter described as "organisms of cells arranged according to a genetic code."

Pythagorean Tree

Problem 9.
Begin with a square.
A) Construct a right isosceles triangle whose hypotenuse is the top edge of the square. Construct squares along each of the other two sides of this isosceles triangle.
B) On the same drawing repeat this construction recursively three times.

Solution:

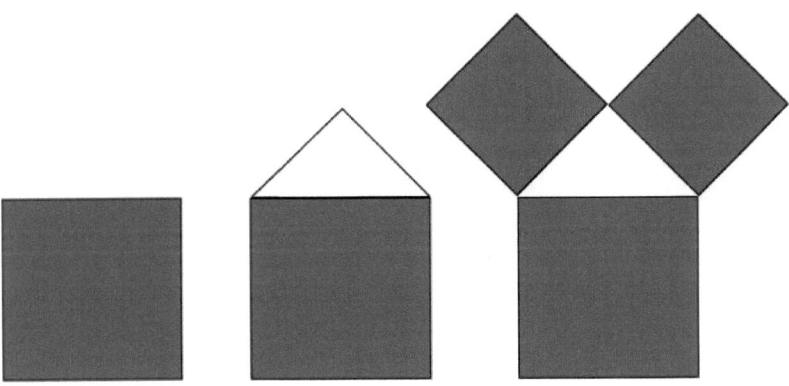

We repeat this construction on each of the two new squares. The figures below show the next two iterations:

The limit of this construction is called the Pythagorean Tree.

The triangles that are attached to each hypotenuse can be any right triangle with acute angles. The image below shows what happens after **20** iterations if each triangle has angles **60°** and **30°**.

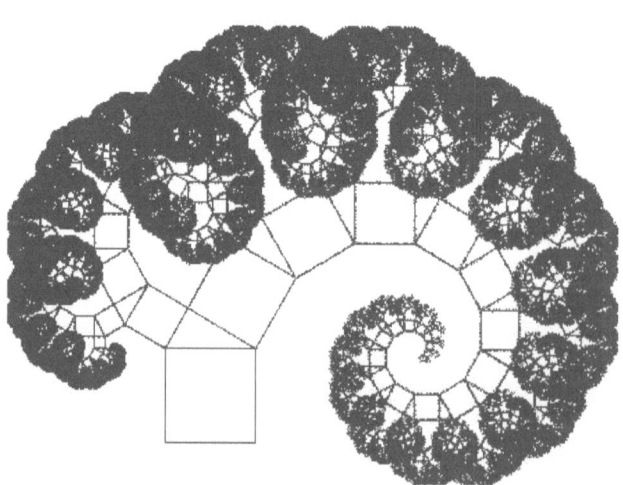

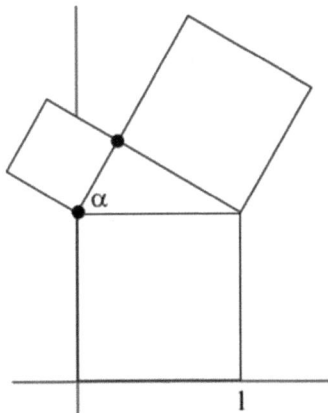

Let's make the initial square a unit square (with side equal to **1**) and with lower left corner at the origin (the beginning of the coordination system). Let **α** be the left angle shown in the figure below.

For the first function, corresponding to the upper left square, we must scale by **cos (α)** and rotate counterclockwise by **α**, then translate straight up. For the second function, corresponding to the upper right square, we must scale by **sin(α)** and rotate clockwise by **90 °-α**, then translate the point at the origin to the point at the right angle in the triangle. Finally, the third function is just the identity. This keeps the squares already drawn in their current locations while the first two functions add additional squares.

$$f1(\mathbf{x})= \begin{bmatrix} cos^2(\alpha) & -cos(\alpha)\sin(\alpha) \\ \cos(\alpha)\sin(\alpha) & cos^2(\alpha) \end{bmatrix}\mathbf{x} + \begin{bmatrix} 0 \\ 1 \end{bmatrix}$$

$$f1(\mathbf{x})= \begin{bmatrix} sin^2(a) & cos(a)\sin(a) \\ -\cos(a)\sin(a) & sin^2(a) \end{bmatrix}\mathbf{x} + \begin{bmatrix} cos^2(a) \\ 1+\cos(a)\sin(a) \end{bmatrix}$$

$$f3(\mathbf{x})= \begin{bmatrix} 1 & 0 \\ 0 & 1 \end{bmatrix}\mathbf{x}$$

In this case the iterations of the initial square under this iterated function system do converge to a limit set. Unlike other examples on this website, however, the limit set will depend on the initial set that is used. The image below shows **10** iterations starting with a square, with **α = 45 °** (the angle of the right triangle). The image on the right shows **10** iterations starting with a vertical line segment.

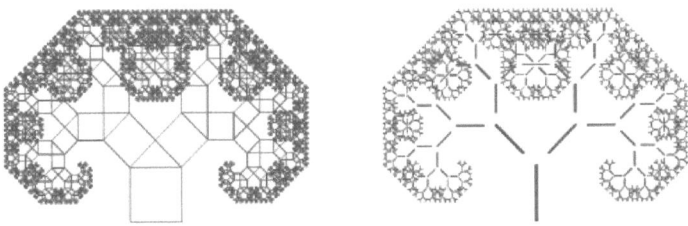

Notice that while the **"trunks"** of the two trees are different, the **"outer leaves"** of the trees do appear to be very similar. This is because if you delete the third

35

function from the iterated function system, the remaining two functions **are** constructive and thus do have a unique attractor no matter what initial set you start with. In fact, functions **1** and **2** are essentially the same functions as for the Lévy Dragon, just with translation vectors shifting an additional **1** unit vertically.

Special Properties

When the Pythagorean tree is drawn with isosceles right triangles ($\alpha = 45$ °) and a unit square (with side equal to **1**) as the initial set, the tree will fit exactly inside a rectangle of width **6** and height **4**. The grid in the image below is **1/2 x 1/2**. The squares will not overlap for the first four iterations, but after that the squares will begin to overlap and start growing back inwards as well as expanding outward to create the "leaf" effect around the perimeter of the tree. The tree will always stay within this rectangle, however.

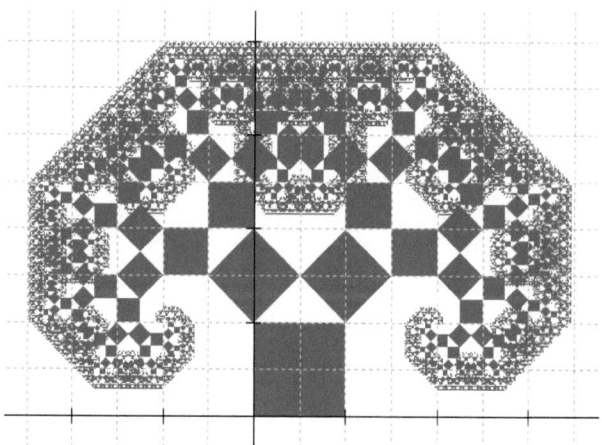

Problem 10.

Let's play with numbers

Take a Pythagorean tree. Suppose you number the squares only. The initial square is labeled with index **1**. If a square has a label **n** and a right isosceles triangle is placed on top of this square, so that two new squares are constructed on it, then the new square on the left will have index **2n** and the new square on the right will have index **2n+1**.

Task: Number the squares on the Pythagorean tree as shown below in **4** iterations.

Solution:

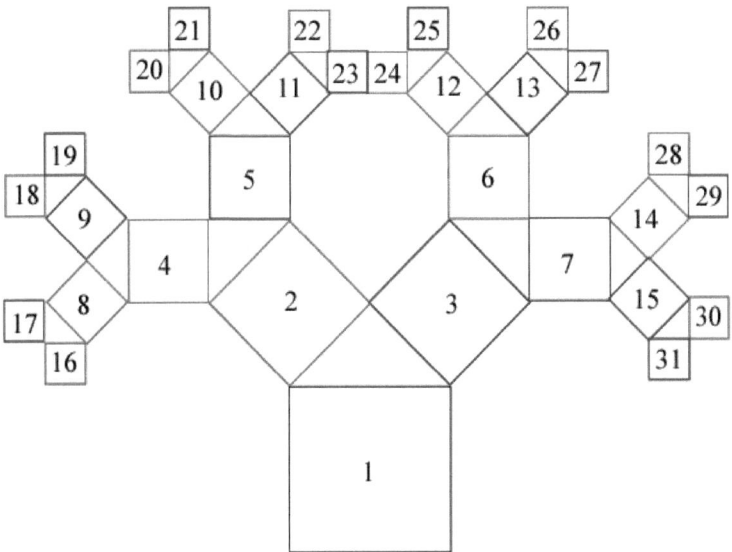

The figure only shows the first **4** iterations. After that the squares will begin to overlap but the indexing algorithm will still apply as more iterations are done. Once a square is given a particular index, it will never change while additional squares are created.

The Pythagorean tree can be used in problems for younger students, like **Problem 1.**

Now we can use it in a problem for older students on the bases of binary notation.

Problem 11.

Take a Pythagorean tree with **4** iterations.

A) Where would square **45** be located?

Solution:

Since **45 = 32 +8 +4 +1,** it would be written in binary as **101101.** The left most digit will always be **1** and corresponds to the base (initial) square. For the remaining digits, a **0** means a turn to the left (**red** square) while **1** means a turn to the right (**blue** square). So to get to square **45**, go left, right, right, left, right. This is illustrated in the figure below.

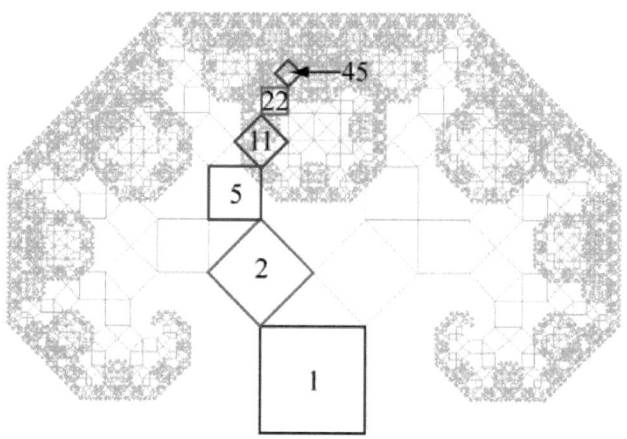

B) Where is square **116**?

Solution:

Since **116 = 64+32+16+4**, in binary this would be **1110100**. So starting at the base, turn right, right, left, right, left, left (which means we need at least **6** iterations to see square **116**).

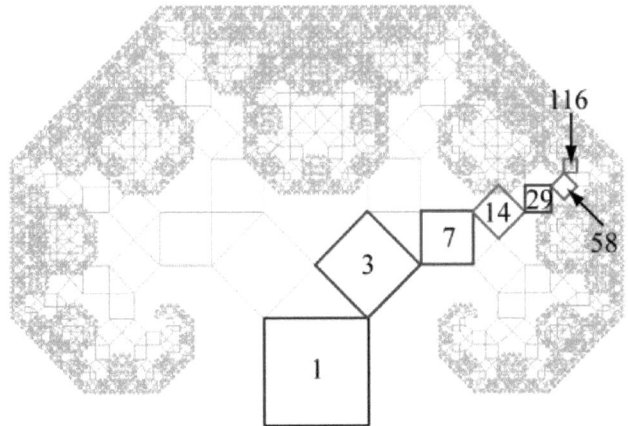

What about square **98156**? In binary this is **10111111101101100** which has **17** digits. You would need to take **16** iterations to get to this square. This square would have been reduced by a factor of $(\sqrt{2} / 2)^{16} = 1/256,$ so it might be a bit too small to see on the computer screen!

The figure on the left below shows the squares with indices **1, 2, 4, 8**, ..., i.e. the powers of **2.** These squares form a logarithmic spiral that converge to a point (more precisely, comparable points on each square lie on a curve that is a logarithmic spiral; the figure on the right shows the curve through the corner points.) In other words, a property of this tree is that its branches follow the path of a logarithmic spiral. But there is really nothing special about starting with the initial square. If you start with any square in the Pythagorean tree and follow the squares off that one that always go either to the left or always go to the right, then those squares will also follow the path of a logarithmic spiral. Thus, we can come to the conclusion that the Pythagorean tree is full of infinitely many spirals.

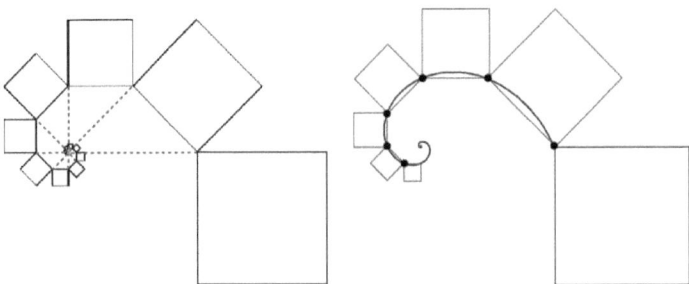

Here is a version of the Pythagorean tree done in back stitch on fabric (14 count per inch). The piece was stitched so that the new squares added at each iteration were done in different shades. Ten iterations are shown, with the last in dark grey to show the development of the Lévy dragon in the very end.

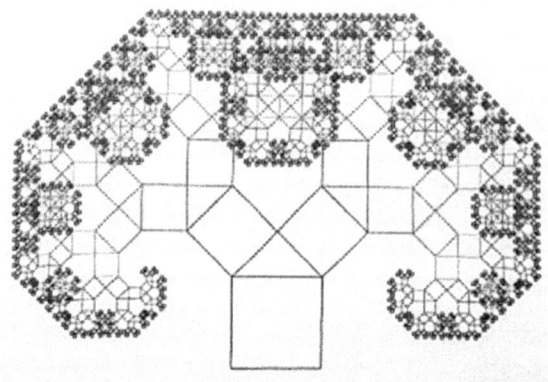

The traditional Pythagorean tree is constructed by starting with a square and constructing two smaller squares such that the corners of the squares coincide pairwise (thus enclosing a right triangle), then iterating the construction on each of the two smaller squares. When viewed as an iterated function system, however, one can start the iteration with any initial image (or set) as in the following image.

For this image we begin with a common picture of Pythagoras as the initial set. The trunk of the three was constructed using **10** iterations of a slight modification of the iterated function system where the first function includes a horizontal reflection across a vertical line. This gives a reflective symmetry to the trunk by having the images of Pythagoras looking towards each other. The picture of Pythagoras was scaled and placed in just the right spot (after some experimentation) so that at each iteration the base of the new pictures will just touch at a **45°** angle. The leaves of the tree consist of **500,000 points** plotted using a random chaos game algorithm and colored based on Michael Barnsley's color stealing algorithm. To give the leaves a more realistic shading, the colors were stolen from a digital photograph of a field of green and yellow grass.

7. Fractal art and other applications of fractals

By studying and constructing fractals we can develop an idea about the processes and phenomena, which are equally complex regardless of their scale. In this way we will be able to help human's understanding of the fractal geometry of nature. It is not exaggerated to say that a co-author of Mandelbrot's discovery is the computer itself. In order to draw a fractal, one has to carry out a large number of computations and show the resultant points on a graph. This is not always possible to be done by hand, whereas the computer is unsurpassed in such operations. It is undeniable that the advent of computer graphics has changed fundamentally the study approach in mathematics and natural sciences. If early scientists relied on the use of numbers and formulas, now scientific studies are much more interesting. With the aid of computers scientists can draw beautiful pictures of the studied phenomena. Some of them have become contemporary artists. Mathematical curiosity towards fractals in the early 1980s has led to a new form of art which nowadays enjoys great popularity – **fractal art**.

Algebra fractals have a significant practical application in computer graphics to create fascinating fractal images.

Fractal geometry has found application in designing antenna devices. This was done for the first time by *Nathan Cohen*, who was living at the time in central Boston. As it was then forbidden to place outdoor antennas on the buildings, Cohen cut a design from aluminum foil in the shape of a Koch's curve, stuck it to a piece of paper and attached it to the TV set. It turned out that such an 'antenna' worked equally well. Although its physical properties were still not studied, Cohen set up his own company and started mass production. Now American company *Fractal Antenna System* has manufactured a fractal antenna of a new generation, which is used in mobile phones.

Problem 12.

Here are 48 cylindrical cans placed in a large box in 8 rows so that each row contains 6 cans. Demonstrate how 50 such cans can be placed in the same box.

Solution: Let's assume that the radius of each can is r. There are 6 cans in each row,

=>The width of the box is at least 6.2r=12r

There are 8 rows => The length of the box is at least 2.8r=16r.

Now, let's arrange the cans as follows:

We put 6 cans in the first row. Next, we put 5 cans in the second row so that each cylinder touches two of the previous row (see picture). The third row is the same as the first; the forth is the same as the second and so forth. So, when we reach the eighth row we will have placed 4(6+5)=44 cans.

We place the rest of the cans – 6 in the ninth row, which makes 50 cans in all.

It is clear that the width of the arrangement is the same (each row of 6 cylinders has a length of 12r, whereas those with 5 cans are shorter in length.

Let's calculate the length of the arrangement:

Each 2 centers in adjacent rows are at symmetrical distance, so let's take a cylinder with center A in the first row and cylinder with center B in the second row, which are adjacent, and define the segment MH. The two cylinders are touching a cylinder with center C.

$MH = \sqrt{3}r$ (triangle ABC is equilateral).

⇨ The total necessary length is $8MH + 2r = 8r\sqrt{3} + 2r$.($2r$ are the intervals from the first and the last row to the outer sides of the box respectively).

$r > 0$,　$49r > 48r \Rightarrow 7r > 4\sqrt{3}r \Rightarrow 14r > 8\sqrt{3}r \Rightarrow 16r > 8\sqrt{3}r + 2r$

=>The necessary length for the new arrangement is smaller compared to that of the previous arrangement.

=> Therefore, we can arrange 50 cans in the same box.

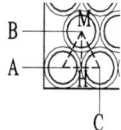

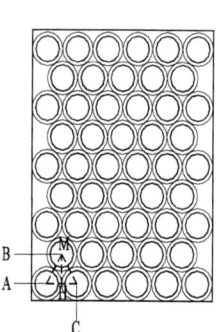

8.Conclusion.

During its evolution, mankind has always strived to get to know the environment it exists in. The knowledge of fractals leads to a better understanding of the laws governing nature and human society. Studying them, we become more aware of the steps and mechanisms of the development of the world around us. In this way, by explaining the chaotic, we can achieve harmony in life Fractal modelling is just the first step towards gaining an understanding of chaos and fractals which surround us. Indeed, by initiating chaos we realize that it is an infinite dimension, which has existed long before this initiation.

By fractal modelling we improve our experience with the unpredictable. Fractals are the **"code"** which puts all seemingly chaotic systems and processes around us into motion.

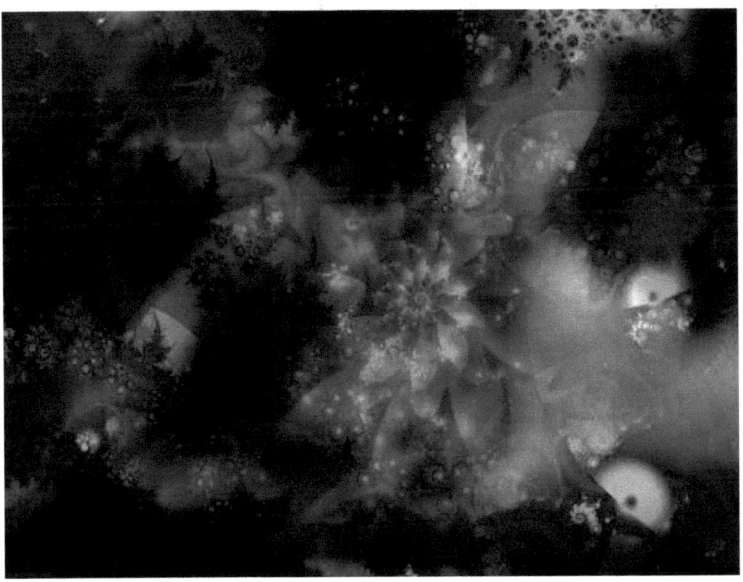

References:

Bogomolny, Alexander. "Koch's Snowflake." Cut The Knot: Learn to Enjoy. http://www.cut-the-knot.org/Curriculum/Calculus/NoLimit.shtml (accessed April 8, 2013).

Frame, Michael, Benoit Mandelbrot, and Nial Neger. "Fractal Geometry." Fractal Geometry. http://classes.yale.edu/fractals/ (accessed January 13, 2013).

К. Славова, Славка, Галина С. Панайотова, and Ивелин Г. Иванов. "ГЕОМЕТРИЧНИ ФРАКТАЛИ В ОБУЧЕНИЕТО НА БЪДЕЩИТЕ НАЧАЛНИ УЧИТЕЛИ." http://fractalmagics.com/. http://fractalmagics.com /files/06_Geometrichni_fraktali.pdf (accessed July 22, 2013).

Mileva, Vanya. "Бръсначът на Окам." Бръсначът на Окам. http://bgchaos.com/ (accessed November 5, 2012).

Nylander, Paul. "Fractals." www.bugman123.com. http://www.bugman123.com/ Fractals/index.html (accessed February 8, 2013).

Po Leung Kuk. "Problem for triangle and square numbers." Hong Kong.

Po Leung Kuk. "Problem for the Sierpinski Triangle." Hong Kong.

Riddle, Larry. "Classic Iterated Function Systems." Agnes Scott College. http://ecademy.agnesscott.edu/~lriddle/ifs/ifs.htm (accessed December 25, 2013).

Suggestion by a team of teachers from Burgas for an international competition MATHEMATICS IN SEARCH OF TALENTS. "Problem with the cans." High School of Mathematics and Natural Sciences „Academician Nikola Obreshkov" Burgas. Publishing House LIBRA SCORP. Burgas 2011

Suggestion by a team of teachers from Burgas for an international competition. "Problem for the Dragon Curve."

Trube, Ben. "Fractals You Can Draw (The Dragon Curve or The Jurassic Fractal)." [BTW] : Ben Trube, Writer. http://bentrubewriter.wordpress.com/2012/04/25/ fractals-you-can-draw-the-dragon-curve-or-the-jurassic-fractal/ (accessed January 21, 2013).

Printed by Books on Demand GmbH, Norderstedt / Germany